# HAWKSBILL TURTLE AS PET

Pet Owners Handbook

A COMPLETE GUIDE TO THEIR CARE AND OWNERSHIP, MAINTENANCE, DIET, HABITAT, AND LOTS MORE

DR MORRIS HART

Copyright© 2024 **DR MORRIS HART**

All rights reserved. No part or part of this book or publication may be reproduced, stored, or transferred in any form by electronic, mechanical, recording, or other retrieval system without written permission from the publisher

# Table of Contents

**INTRODUCTION** ................................................................. 5

**CHAPTER 1** ........................................................................ 7

The Appropriate Hawksbill Turtle Selection ..................... 7

**CHAPTER 2** ...................................................................... 16

Creating the Perfect Environment for Your Hawksbill Turtle ................................................................................. 16

**CHAPTER 3** ...................................................................... 26

The Nutrition and Feeding Needs of Hawksbill Turtles .. 26

**CHAPTER 4** ...................................................................... 34

Guidelines for the Care and Management of Hawksbill Turtles ................................................................................ 34

**CHAPTER 5** ...................................................................... 43

Medical Treatment and Typical Problems for Hawksbill Turtles ................................................................................ 43

**CHAPTER 6** ............................................................. **54**

ADVANCED CARE ADVICE FOR SKILLED HAWKSBILL TURTLE KEEPERS.................................................................. 54

**CHAPTER 7** ............................................................. **65**

LEGAL AND MORAL ISSUES WITH HAWKSBILL TURTLE RESIDENCY ............................................................................ 65

**CHAPTER 8** ............................................................. **75**

FAQS: COMMON QUESTIONS ADDRESSED CONCERNING HAWKSBILL TURTLES..................................................... 75

**CHAPTER 9** ............................................................. **85**

FINAL THOUGHTS: SAVORING YOUR HAWKSBILL TURTLE FRIEND 85

# Introduction

Scientifically known as Eretmochelys imbricata, hawksbill turtles are amazing animals that have won over the hearts of many reptile enthusiasts. Because of their remarkable appearance and distinctive habits, hawksbill turtles are valued for their important role in marine ecosystems in addition to their beauty.

Hawksbill turtles are found in tropical and subtropical regions of the world. They are well-known for their beautiful shell designs, which are made up of overlapping scales that resemble roof tiles. Hawksbill turtles live mostly in coral reefs, where their ability to control the populations of other invertebrates and feed on sponges is essential to the health of these fragile ecosystems.

Although Hawksbill turtles are best known for living in the wild, some are kept as pets by passionate reptile enthusiasts. Taking care of a Hawksbill turtle demands a lot of thought and dedication because they have particular needs when it comes to their habitat, food, and socialization that need to be satisfied for them to be healthy and happy.

Whether you're a seasoned reptile keeper or a novice to the world of turtle care, this comprehensive resource will offer insightful information and useful tips for creating a rewarding and enriching environment for your Hawksbill turtle companion. We will cover everything you need to know about keeping Hawksbill turtles as pets, from choosing the right turtle to providing proper housing, nutrition, and healthcare.

# Chapter 1

## The Appropriate Hawksbill Turtle Selection

Making wise choices is crucial when starting the process of obtaining a Hawksbill turtle as a pet, as it will protect the turtle and you as the caregiver. Choosing the appropriate turtle requires careful consideration of a number of factors, such as the turtle's origin, health, and unique characteristics. In this section, we'll go over the important factors to take into account when choosing a Hawksbill turtle as a pet.

- Knowledge of Hawksbill Turtles

A thorough understanding of the natural history, behavior, and habitat requirements of Hawksbill turtles is essential before beginning the selection process. Hawksbill turtles are medium-sized sea turtles that are globally found in tropical and subtropical oceans, with a special preference for coral reef habitats. They are distinguished by their distinctive beak-like jaws and intricately patterned shells.

Hawksbill turtles are mostly carnivorous in the wild, however they also eat algae and other invertebrates. They are essential to the health of coral reefs because they regulate sponge populations, which can overgrow and suffocate coral if left unchecked.

Though they can adapt to captivity under the right circumstances, it is imperative that prospective pet owners understand that Hawksbill turtles have unique needs and behaviors shaped by their natural environment. A close replica of their natural habitat is also necessary for their physical and psychological well-being.

- Things to Think About Before Purchasing a Hawksbill Turtle

A potential owner should carefully assess their level of preparedness and commitment to caring for Hawksbill turtles before obtaining one. Maintaining a Hawksbill turtle is a long-term commitment that calls for sufficient funds, time, and attention. Take into account the following factors:

Legal Considerations: Make sure you are in accordance with all applicable laws and, if needed, obtain the appropriate permits or licenses. Research local and international rules pertaining to the ownership and importation of Hawksbill turtles.

Space Requirements: Make sure you have enough space to fit a good habitat for your Hawksbill turtle. These creatures need large enclosures with access to both land and water.

Be ready for the financial commitment that comes with owning a Hawksbill turtle, as maintaining one requires regular costs for food, materials, veterinarian care, and habitat setup.

Time and Effort: Think about if you have the time and energy to commit to the regular care and maintenance that hawksbill turtles require, such as feeding, cleaning, and health monitoring.

Lifespan: In captivity, Hawksbill turtles can live for several decades, with some individuals reaching 50 years or more. Adopting a Hawksbill turtle requires a long-term commitment.

Compatibility with Other Pets: Take into account if your Hawksbill turtle will get along well with any other pets you may already have in your home, such as fish, reptiles, or other turtles.

- How to Choose a Healthful Hawksbill Turtle

To guarantee a happy beginning for your turtle ownership experience, pick a healthy Hawksbill turtle. Look for the following indicators of health and vitality:

Alertness: A healthy Hawksbill turtle should have bright eyes, a robust, energetic personality, and be attentive and responsive.

Shell Condition: A healthy shell should be smooth, firm, and devoid of any abnormalities. Inspect the turtle's shell for any indications of damage, such as chips, cracks, or soft patches.

Body Condition: The turtle should have a well-rounded body form and a healthy weight. Look for symptoms of malnutrition or disease, such as emaciation, lethargy, or abnormal swelling.

Skin and Eyes: Check for lesions, ulcers, and unusual pigmentation on the turtle's skin. Make sure its eyes are clear and unclouded.

Respiratory Function: Pay attention to the turtle's respiration for any indications of respiratory distress, like gasping, wheezing, or difficult breathing.

- Origin of Purchase

Take into consideration the following choices for obtaining a Hawksbill turtle: It is important to select a reliable provider in order to guarantee the health, welfare, and legal status of the turtle.

Licensed Breeders: Make sure the breeder has the necessary qualifications and ask about the turtle's health history, genealogy, and any accessible documents. Only buy Hawksbill turtles from licensed breeders who follow ethical breeding techniques and provide their turtles the care they need.

Rescue Organizations: If you're looking to adopt a Hawksbill turtle, there are plenty of respectable rescue groups and animal shelters that specialize in taking care of and finding new homes for turtles in need.

Legal Channels: Make sure that the purchase of your Hawksbill turtle conforms with all applicable rules and regulations that control the ownership and trading of protected species. Refrain from buying turtles from sources that are prohibited by law or that are unethical, including unlicensed dealers or wildlife traffickers.

- Personal Aspects

Every Hawksbill turtle has an individual personality, set of preferences, and set of characteristics. Take into account the following while choosing a turtle:

Size: Adult Hawksbill turtles normally weigh between 100 and 200 pounds and range in length from 2 to 3 feet. Select a turtle size based on the amount of space you have available and your capacity to give proper care.

Age: When choosing a turtle, take into account its age. Younger turtles might need more regular care and observation, while older turtles might need special attention to their health.

Gender: Hawksbill turtles vary in size, behavior, and reproductive structure between male and female individuals. If at all possible, ascertain the turtle's gender since this may impact its behavior and care needs.

Compatibility: Take into account how well the turtle will get along with your home, way of life, and other pets. While some turtles do better in social settings with other turtles, others may prefer to live alone.

Personal Connection: Spend some time getting to know the turtles up for adoption and select one with whom you have a strong emotional connection. Developing a strong bond with your adopted turtle is crucial to a happy and satisfying relationship.

You can start your rewarding journey of turtle ownership with confidence and enthusiasm by carefully weighing these factors and taking your time in choosing the ideal Hawksbill turtle for your home and lifestyle. Just keep in mind that proper care and attention is essential to guaranteeing the health and happiness of your pet turtle for many years to come.

# Chapter 2

## Creating the Perfect Environment for Your Hawksbill Turtle

Because Hawksbill turtles live in coral reef ecosystems in the wild, it's critical to create an environment as similar to this one as possible in captivity. In this extensive guide, we'll go over the important elements and factors to consider when setting up the perfect habitat for your Hawksbill turtle, such as enclosure selection, environmental parameters, substrate options, basking and swimming areas, lighting and heating requirements, and enrichment opportunities.

- Choice of Enclosure:

Choosing a suitable enclosure is the first stage in preparing your habitat for a Hawksbill turtle. Because of their terrestrial and aquatic activities, Hawksbill turtles need large spaces with both land and water regions. When choosing an enclosure, take into account the following factors:

Size: Adult Hawksbill turtles reach a maximum length of 2 to 3 feet, making them medium-sized reptiles. Make sure your habitat has enough space for swimming, diving, and sunbathing. For aquatic turtles, you should allow at least 10 gallons of water for every inch of shell length.

Material: Make sure the cage is escape-proof and strong enough to support the turtle's weight and activity. Use materials like glass, acrylic, or high-density polyethylene (HDPE) that are non-toxic and long-lasting.

Accessibility: Take into account the height of the enclosure walls to prevent your turtle from trying to escape, and choose an enclosure with easily accessible doors or openings for simple cleaning, maintenance, and interaction with your pet.

Customization: Build a naturalistic and enriching environment that replicates the Hawksbill turtle's natural habitat by incorporating ramps, platforms, hiding places, and plants to suit your turtle's individual needs.

- Environmental Factors:

Your Hawksbill turtle's health and well-being depend on maintaining proper environmental conditions. Keep an eye on and adjust the following to establish a cozy and stable habitat:

Temperature: To ensure optimal health and well-being, hawksbill turtles need a basking area that is between 85 and 90°F (29 and 32°C) with water that is between 75 and 80°F (24 and 27°C). Thermometers should be used to track temperature variations and make necessary adjustments to the heating sources.

Humidity: To avoid respiratory problems and dehydration, keep the enclosure's relative humidity between 60% and 80%. Hygrometers can be used to measure humidity levels, and regular misting or humidification can be applied as needed.

Water Quality: To maintain a healthy aquatic habitat for your turtle, make sure the enclosure's water is clean, filtered, and free of pollutants. You should also regularly check water quality factors including pH, ammonia, nitrite, and nitrate levels.

Air Quality: Use fans or ventilation devices to enhance air circulation and exchange inside the habitat. Enough ventilation should be provided within the enclosure to prevent the accumulation of stagnant air and respiratory problems.

- Options for Substrates:

The following substrate options should be taken into consideration for the enclosure's land and water portions in order to provide a clean and pleasant habitat for your Hawksbill turtle:

Aquatic Substrate: For the bottom of the water area, use a substrate material such as sand, aquarium gravel, or large river rocks; steer clear of substrates that the turtle could ingest and cause impaction.

Land Substrate: For the land portion of the enclosure, use a substrate material like as coconut coir, cypress mulch, or reptile-safe soil. Allow for burrowing and nesting behaviors by providing a minimum substrate depth of 4 to 6 inches.

Cleaning and Maintenance: Spot-clean the land area as needed, and use a siphon or gravel vacuum to remove material from the water area to prevent the accumulation of waste and bacteria. Clean and replace the substrate on a regular basis.

- Swimming and Sunbathing Areas:

To suit their thermoregulatory and behavioral demands, hawksbill turtles must have access to both swimming and basking places. The enclosure should be designed with the following areas included:

Basking Area: Whether it's flat rocks, driftwood, or commercial basking platforms, make sure the surface is firm and dry so your turtle may emerge from the water to relax and thermoregulate.

UVB Lighting: To give your turtle the UVB radiation it needs for calcium metabolism and vitamin D synthesis, install a UVB lamp above the basking area. Make sure the UVB lamp is made specifically for reptiles, and replace or relocate the bulb according to the manufacturer's instructions.

Swimming Area: Make sure the water depth permits your turtle to fully submerge and move around freely. Use aquarium filters and water heaters to maintain water quality and temperature. Provide a big, deep water area for your turtle to swim, dive, and explore.

- Needs for Heating and Lighting:

The health, metabolism, and behavior of your Hawksbill turtle depend on proper lighting and heating. Install the following lighting and heating components in the enclosure:

Basking Light: Position the basking light above the basking area and use a timer to replicate a natural day-night cycle. Use a heat lamp or ceramic heat emitter to produce a basking spot with temperatures between 85 and 90°F (29 and 32°C).

UVB Radiation: To provide your turtle the UVB radiation it needs for calcium metabolism and vitamin D synthesis, install a UVB lamp with an acceptable UVB output. Position the UVB lamp above the basking area and make sure it is lit for the recommended amount of time and intensity according to the manufacturer's instructions.

Heating Elements: To maintain ideal circumstances, keep the swimming area's water temperature between 75 and 80°F (24 and 27°C) by using submersible water heaters or aquarium heaters. Regularly check the water temperature and make necessary adjustments to the heating sources.

- Opportunities for Enrichment:

Incorporate the following enrichment components into the cage to encourage your Hawksbill turtle to engage in natural behaviors and to maintain physical and mental health:

Hiding Spots: Whether on land or in the water, use rocks, driftwood, PVC pipes, or commercial hideouts to create hiding spots and sheltered areas where your turtle can retreat and feel safe.

Environmental Enrichment: To create a visually pleasing and exciting habitat, add natural decor items like real or fake plants, branches, rocks, and shells. Arrange the decor so that your turtle has obstacles, climbing chances, and exploration places.

eating Enrichment: Promote your turtle to explore and participate in natural eating habits to maintain physical and mental stimulation. Offer opportunities for foraging and hunting by providing live prey, puzzle feeders, or floating food items.

Maintaining optimal health and well-being for your pet turtle requires regular monitoring and habitat condition adjustments. By carefully planning and setting up the ideal habitat for your Hawksbill turtle, you can create a safe, comfortable, and enriching environment where your turtle can thrive and flourish for years to come.

# Chapter 3

## The Nutrition and Feeding Needs of Hawksbill Turtles

Since they are omnivores, Hawksbill turtles have a varied diet that consists of a variety of plant and animal matter. In this extensive guide, we'll go over feeding practices, nutritional needs, dietary requirements, frequency of feedings, and additional considerations for Hawksbill turtles to make sure they receive a balanced and nutritious diet.

- Comprehending the Eating Behavior of Hawksbill Turtles:

As opportunistic feeders, hawksbill turtles will eat a wide variety of prey items found in their natural environment. In the wild, they mostly eat sponges, but they also eat other invertebrates, algae, and plant matter. Their diet varies depending on their age, habitat, and the seasonal availability of food resources.

To maintain good nutrition and general health, Hawksbill turtles kept in captivity need a diet that closely mimics their natural eating habits. Providing a varied and balanced food is vital for satisfying their nutritional needs and fostering overall well-being.

- Needs for Nutrition in Hawksbill Turtles:

To sustain their growth, development, and physiological processes, hawksbill turtles have particular nutritional requirements that must be satisfied. Make sure your turtle's diet contains the following vital nutrients:

Provide sources of animal protein such as fish, shrimp, insects, and mollusks. Protein: To support muscle development, tissue repair, and metabolic activities, hawksbill turtles need a diet rich in protein.

Fiber: Include fibrous plant materials such as leafy greens, vegetables, and fruits in your Hawksbill turtle's diet to give roughage and aid in digestion. Fiber is crucial for preserving digestive health and fostering healthy gut function in these turtles.

Vitamins and Minerals: Provide a wide variety of foods high in vitamins A, D, and E, as well as calcium, phosphorus, and other important minerals. Hawksbill turtles need a variety of vitamins and minerals to support their overall health and immunological function.

Calcium: Provide sources of calcium such as leafy greens, calcium supplements, and cuttlebones to guarantee proper consumption. Calcium is especially vital for Hawksbill turtles to support shell growth, bone density, and muscular function.

Provide sources of omega-3 fatty acids, such as fish and seafood, to Hawksbill turtles. Omega-3 fatty acids are important for the health of these turtles, especially for preserving the integrity of their skin and shell, lowering inflammation, and promoting cardiovascular health.

- The Nutritional Needs of Hawksbill Turtles:

To ensure that your Hawksbill turtle is getting enough nourishment, make sure their diet is well-balanced and varied. Here are some dietary requirements and tips to keep in mind:

Plant Matter: Give your turtle a diet rich in vitamins, minerals, and fiber by offering leafy greens, vegetables, and fruits like kale, collard greens, spinach, carrots, squash, bell peppers, and berries.

Animal Protein: Feed your turtle a variety of fish, shrimp, crickets, mealworms, earthworms, and other insects to boost muscle growth and metabolic processes. This will supply a broad spectrum of important amino acids.

Commercial meals: Choose high-quality commercial meals that offer a balanced combination of nutrients and vitamins to complement fresh foods. You may supplement your turtle's diet with commercial turtle pellets or blocks that are specifically developed for aquatic turtles.

Feeding Frequency: To replicate their natural eating habits and aid in digestion, feed your Hawksbill turtle short, frequent meals throughout the day. Make sure each feeding session includes a mixture of plant and animal matter for a well-balanced diet.

Variety is key to preventing dietary inadequacies and promoting nutritional diversity. To keep your turtle interested in eating, rotate and alternate between different types of food.

- Tips and Considerations for Feeding:

To ensure a secure and pleasurable feeding experience, take into account the following advice and factors when feeding your Hawksbill turtle:

Provide Appropriate Portion Sizes Based on Your Turtle's Age, Size, and Activity Level: To prevent obesity and associated health problems, don't overfeed your turtle and keep an eye on portion sizes.

Prey Size: Cut larger food items into smaller bits or slices to make them easier for your turtle to eat. Provide prey items that are suitable for the size of your turtle's mouth and digestive system.

Calcium Supplementation: Dust prey items, veggies, or fruits with calcium powder before feeding to give them an extra calcium boost. Supplement your turtle's diet with calcium powder or calcium-rich meals to guarantee optimal consumption.

Gut Loading: To maximize the nutritional value of live prey items and guarantee your turtle gets the most nutrients possible, make sure the insects, for example, are appropriately gut-loaded with meals before putting them in your diet.

Hydration: To avoid dehydration and encourage appropriate hydration, make sure your turtle always has access to clean, freshwater. You should also provide a shallow water dish or tray inside the enclosure for it to sip from and soak in.

Observation: Depending on your turtle's unique tastes and nutritional needs, you may need to make adjustments to their diet and feeding schedule. Regularly observe your turtle's eating behavior, appetite, and general health to spot any changes or anomalies.

With proper care and attention to dietary requirements, you can ensure a long and healthy life for your pet Hawksbill turtle. By following these feeding and nutrition guidelines, you can provide your Hawksbill turtle with a balanced and nutritious diet that supports their health, growth, and vitality. Don't forget to consult with a reptile veterinarian or experienced herpetologist for personalized dietary recommendations and advice specific to your turtle's needs.

# Chapter 4

## Guidelines for the Care and Management of Hawksbill Turtles

The handling, bonding, and safety precautions of Hawksbill turtles are all covered in this extensive guide. Hawksbill turtles are fascinating and captivating animals that can make wonderful pets for reptile enthusiasts. However, it's important to understand how to handle and interact with these turtles safely and responsibly to ensure their well-being and minimize stress.

- Recognizing the Behavior of Hawksbill Turtles:

Hawksbill turtles are shy, solitary creatures that prefer to spend most of their time in their natural habitat. When kept in captivity, they may exhibit varying degrees of tolerance to handling and interaction, depending on their individual personalities and prior experiences. It is important to understand a Hawksbill turtle's behavior and temperament before attempting to handle or interact with one.

When interacting with your turtle, it's important to respect its preferences and boundaries and to watch for signals of stress or discomfort through body language. If your turtle exhibits aggression, fear, or retreat, don't force interactions or handling.

- Managing Methods:

To reduce stress and guarantee the safety of the handler and the turtle, it is imperative to employ gentle and polite handling procedures. The following rules should be followed:

Approach quietly: To prevent shocking or startling your turtle, approach them quietly and slowly. Give them time to get used to your presence before trying to touch or handle them.

Support the Body: To avoid hurting your turtle, avoid gripping or squeezing them tightly as this may cause discomfort or suffering. Instead, support the body of your turtle with both hands when raising or holding them.

Lift from Below: To assist your turtle feel more safe and less stressed while handling, lift them from below instead of grabbing them from above. This will replicate the feeling of being supported in the water.

Avoid Sudden Movements: To keep the ambiance peaceful and relaxed, steer clear of making abrupt or jerky movements that could frighten or upset your turtle. Instead, walk gently and gradually.

Respect Boundaries: When your turtle exhibits indications of stress or discomfort, let them go to their hiding place or basking area; do not handle them excessively or for extended periods of time.

- Guidelines for Interaction:

Even though Hawksbill turtles might not be as rigorous as other creatures, you can still establish a relationship with them by following these recommendations for interaction:

Observation: Take some time to watch your turtle from a distance to get a sense of their preferences, routines, and personalities. Then, use this information to customize your interactions and offer enrichment that piques their interests.

Respect Personal Space: Give your turtle the freedom to explore and engage with their surroundings at their own pace. Do not trespass on their territory or interfere with their routine in any way.

Hand Feeding: Use tweezers or tongs to deliver little pieces of food, such as leafy greens or insects, and let your turtle approach and eat at their own pace. This approach fosters trust and pleasant associations with your presence.

Enrichment Activities: Give your turtle toys and activities that will encourage natural behaviors and pique their curiosity. You can give them puzzle feeders, floating objects, or interactive toys to play with.

Supervised Exploration: Keep a constant eye on your turtle to protect their safety and prevent damage or escape. Allow your turtle to explore outside of its habitat in a safe and supervised location, such as a dedicated play area or secure outdoor enclosure.

- Bonding Techniques:

Patience, consistency, and mutual trust are necessary while connecting with your Hawksbill turtle. The following bonding techniques can help you and your turtle grow closer over time:

Take Time to Spend Quality Time with Your Turtle: Schedule regular time each day to spend with your turtle, feeding, exploring, or just watching. Creating a pattern will make your turtle feel more at ease and confident in your company.

Positive Reinforcement: Reward desired behavior and promote cooperation and trust by using strategies like praise or treats. Reward calm and relaxed behavior and refrain from using negative reinforcement or punishment.

Respect Individual Preferences: Take into account and honor each turtle's unique preferences and personality traits. Certain turtles may be more comfortable being handled and handled physically, while others may prefer to watch from a distance. Adjust your interactions to suit your turtle's needs.

Be Patient and Understanding: Developing a relationship with your turtle requires time and patience, so as you build rapport and trust, be patient and understanding. Don't rush or push conversations; instead, give your turtle the space and time they need to become confident and at ease.

- Safety Measures:

Prioritizing the safety of both the handler and the turtle is crucial when working with Hawksbill turtles. Take the following safety measures:

Wash Your Hands: To stop the transmission of bacteria and lower the chance of infection, wash your hands well both before and after handling your turtle.

Supervise Interactions: Keep small children and other animals away from the turtle's enclosure unless under direct observation. Always supervise interactions between your turtle and other pets or family members to prevent accidents or injury.

Handle with Care: Try not to hold or constrain your turtle with force since this might trigger defensive or aggressive responses. Instead, handle your turtle gently and carefully to prevent harm or stress.

Avoid physical Play: To keep a pleasant and trusting connection with your turtle, avoid physical play or handling that could frighten or upset them. Instead, respect their comfort level and boundaries.

With a little time and effort, you can create a lifetime bond with your pet turtle that is meaningful and strong. By adhering to these handling and interaction guidelines, you can guarantee a safe, enjoyable, and rewarding experience for both you and your Hawksbill turtle. Always remember to approach interactions with patience, respect, and understanding.

# Chapter 5

## Medical Treatment and Typical Problems for Hawksbill Turtles

As with all animals, Hawksbill turtles can suffer from minor ailments to more serious conditions, so it's important to ensure their health and well-being to ensure their longevity and quality of life. In this extensive guide, we'll go over health care procedures, preventative measures, common health issues, symptoms to look out for, and what to do in the event of an illness or injury.

- Standard Medical Procedures:

To keep your Hawksbill turtle in top shape, you must maintain routine health care practices. To help your turtle stay healthy and happy, do the following routine care practices:

Habitat Maintenance: To keep waste, bacteria, and parasites from accumulating in your turtle's habitat, keep it clean and well-maintained. Regularly change the water in the enclosure, clean the substrate, and sanitize it.

Observation: Keep a journal of your turtle's routines and overall health to follow any changes or trends over time. Regularly check in on your turtle's behavior, appetite, and appearance to spot any anomalies or changes.

Nutrition: Offer a range of fresh meals, supplements, and commercial diets to guarantee appropriate intake of vital nutrients and vitamins. A balanced and nutritious diet that suits your turtle's individual dietary demands should be provided.

Environmental Parameters: Use thermometers, hygrometers, and water testing kits to monitor environmental conditions and make necessary adjustments. Maintain the enclosure's proper temperature, humidity, and water quality to support your turtle's health and physiological functions.

Veterinary Check-ups: Make sure your turtle receives routine veterinary check-ups to determine its general health, identify any underlying problems, and administer preventive care, such as vaccinations and parasite screenings. For individualized treatment and advice, speak with a veterinarian who specializes in treating Hawksbill turtles.

- Preventive actions:

Maintaining the health and well-being of your Hawksbill turtle requires proactive health issues prevention. Include the following preventative actions in your turtle care regimen:

Before adding any new turtles or animals to your collection, quarantine them to stop the spread of parasites and diseases. During the quarantine period, keep a watchful eye out for any signs of infection or illness in any new additions.

Proper Handling: Use gentle handling techniques and minimize physical contact to lessen the chance of damage or stress-related health issues. Treat your turtle with care and avoid needless stress or trauma that could compromise their immune system.

Live plants, décor pieces, and substrates should all be quarantined before being added to your turtle's enclosure in order to avoid introducing pests, infections, or pollutants. Before adding anything into the habitat, make sure everything is well inspected and cleaned.

Water Quality Management: Use filtration systems, water conditioners, and routine water changes to ensure clean and healthy water for your turtle. Regularly test and monitor water parameters, such as pH, ammonia, nitrite, and nitrate levels.

In order to prevent and manage common external and internal parasites that may affect Hawksbill turtles, implement a parasite control program. Speak with your veterinarian to create a parasite prevention plan that is specific to your turtle's needs and risk factors.

- Typical Health Concerns:

Hawksbill turtles may occasionally have health problems despite your best efforts to give them the care and preventative measures they need. Be familiar with the following common health conditions and their signs so you can identify and treat any concerns as soon as possible:

Wheezing, gasping, nasal discharge, and lethargy are some of the symptoms of respiratory infections, which are frequent in Hawksbill turtles and can be brought on by bacterial and fungal pathogens, poor water quality, or environmental stress.

The fungal or bacterial ailment known as shell rot can cause soft areas, discoloration, foul odor, and shell erosion in Hawksbill turtles. It is usually brought on by injuries, poor hygiene, or environmental stress.

Metabolic Bone Disease (MBD): Softening or distortion of the shell, limb weakness, trouble walking, and stunted growth are some of the symptoms of this nutritional insufficiency that affects Hawksbill turtles, namely calcium and vitamin D3 shortages.

Parasitic Infections: Depending on the type and severity of the parasite infestation, symptoms such as weight loss, lethargy, decreased appetite, and abnormal behavior may occur. Hawksbill turtles may be vulnerable to internal and external parasites, such as nematodes, cestodes, and ectoparasites.

Eye Issues: Infections, traumas, or inflammations can impair the vision and general health of hawksbill turtles. Symptoms may include discharge, redness, swelling, or cloudiness of the eyes, as well as difficulties opening or shutting them.

- Identifying Symptoms:

For your Hawksbill turtle to receive appropriate veterinary care and treatment, it is imperative that any illness or injury symptoms are promptly identified and addressed. Keep a close eye out for any of the following signs and symptoms that could point to a health problem for your turtle:

Appetite Changes: Reduced appetite, aversion to food, or adjustments to feeding habits may be signs of underlying medical conditions, digestive disorders, or nutritional deficiencies.

Lethargy: If your turtle exhibits increased weakness, lethargy, or unwillingness to move, it may be a sign of underlying health problems, stress, or environmental difficulties that are hurting its overall wellbeing.

Abnormal Behavior: Any behavioral changes in your turtle, such as odd swimming patterns, excessive hiding, or hostility, could be an indication of stress, disease, or discomfort.

Wheezing, gasping, coughing, or nasal discharge are examples of respiratory symptoms that may point to respiratory infections or other respiratory issues that need to be treated by a veterinarian.

Physical Abnormalities: A veterinarian should examine any physical abnormalities or changes in appearance, such as discolouration, abnormal growths, lesions on the shell, or deformities in the shell, to ascertain the underlying cause and the best course of action.

- What to Do If it Gets Sick or Get Hurt:

It's critical to respond quickly to address your Hawksbill turtle's health concerns if you see any indications of illness or injury. In the event of an illness or injury, follow these steps:

Isolate the Turtle: To stop the disease from spreading or harming other turtles, take the afflicted turtle out of the main cage and place it in a different quarantine tank or container.

Offer Supportive Care: The afflicted turtle should get supportive care, which includes preserving ideal environmental conditions, providing wholesome food, and reducing stressors that could make their condition worse.

Seek Veterinary Care: Make an appointment with a veterinarian who specializes in treating Hawksbill turtles as soon as possible to have your turtle examined by one. You should also bring along a thorough history of your turtle's symptoms, behavior, and habitat conditions to help with diagnosis and treatment planning.

Treatment Recommendations: To effectively address your turtle's health issues, carefully follow your veterinarian's treatment recommendations and instructions. Administer drugs, vitamins, or other therapies as indicated, and keep a close eye on your turtle's response to treatment.

Keep an Eye on Progress: Keep a close eye on your turtle's development and reaction to treatment, and promptly report any changes or worries to your veterinarian. Make sure your turtle receives any suggested follow-up appointments or diagnostic tests to guarantee its sustained healing and wellbeing.

Remember to prioritize regular monitoring, prompt intervention, and veterinary care to address any health issues promptly and ensure a long and healthy life for your pet turtle. By putting these health care practices, preventative measures, and response strategies into practice, you can help safeguard the health and well-being of your Hawksbill turtle and give them the best care possible throughout their life.

# Chapter 6

## Advanced Care Advice for Skilled Hawksbill Turtle Keepers

Expert Hawksbill turtle caretakers can improve their turtles' quality of life and overall well-being by going above and beyond basic care. There are various approaches to improve the care given to Hawksbill turtles in captivity, ranging from sophisticated habitat design to specialist medical procedures. We'll go over advanced care methods and advice in this extensive guide for seasoned keepers who want to improve the quality of their turtle care.

- Superior Design for Habitat:

The design of an advanced habitat for Hawksbill turtles must take into account their natural activities, preferences for their surroundings, and need for enrichment. Elements of an advanced habitat design could be:

A naturalistic enclosure should be designed with elements like driftwood, live plants, rock formations, and surfaces made of sand or gravel to resemble the habitat of Hawksbill turtles. Construct a multifaceted, diverse terrain setting for learning and growth.

Bioactive Setup: To construct a self-sustaining environment, think about putting living plants, helpful microorganisms, and realistic substrate layers into the turtle enclosure. In addition to lowering maintenance needs and offering turtles enrichment, bioactive installations can aid in maintaining ideal environmental conditions.

Aquascaping: Use rock, driftwood, and aquatic plant arrangements to create caves, ledges, and hiding places for the turtles. This will make an aesthetically pleasing underwater landscape inside the enclosure. Create a dynamic and visually appealing aquatic environment by utilizing natural materials and landscaping equipment.

Personalized Enrichment: Offer your Hawksbill turtles individualized enrichment experiences based on their unique tastes and behaviors. To pique their interest and promote natural behaviors, provide them with an assortment of interactive toys, puzzle feeders, floating platforms, and climbing structures.

- Advanced Techniques for Feeding:

Hawksbill turtles are fed using advanced techniques that include varying their food, adding live prey, and maximizing nutritional supplements. Take into account these sophisticated feeding methods:

Gut Loading: Before presenting live prey to your turtles, such as insects and feeder fish, pre-load them with nutrient-rich meals. Gut loading increases the nutritional content of prey items and gives the turtles another supply of vitamins and minerals.

Target Feeding: To guarantee that every turtle in a multi-species cage gets an equal amount of food, use target feeding techniques. To enable individual feeding sessions, teach the turtles to identify and react to feeding cues, such as a particular feeding station or target stick.

Seasonal Variability: Modify the turtle's diet according to the changing needs for nutrition and variations in the availability of natural food. Provide a wide variety of foods, including as in-season fruits, vegetables, and prey items, to give their diet year-round diversity and balance.

Supplement Rotation: To guarantee thorough supplementation and avoid overdosing or inadequacies, alternate between several kinds of nutritional supplements, such as calcium powders, vitamin supplements, and mineral blocks. Observe the recommended dosage and frequency of supplementation as stated by the manufacturer.

- Advanced Medical Procedures:

Hawksbill turtles receive advanced medical care that includes specialized therapies, diagnostic testing, and proactive monitoring to address complicated health conditions. Think about the following cutting-edge medical procedures:

Regular Health evaluations: To identify any early indicators of sickness or anomalies, do routine health evaluations on your Hawksbill turtles. These assessments should include physical checks, weight tracking, and fecal analysis. For reference and tracking, keep thorough records of your turtle's medical history and current state.

Diagnostic Imaging: Assess internal structures, identify anomalies, and identify underlying medical issues in Hawksbill turtles by using diagnostic imaging modalities like radiography (X-rays) and ultrasound. For an accurate diagnosis and treatment planning, speak with a veterinarian who specializes in diagnostic imaging interpretation for reptiles.

Blood Testing: To evaluate the general health, organ function, and nutritional condition of your turtle, perform normal blood tests, such as complete blood counts (CBC) and blood chemistry panels. Blood testing can help inform treatment choices and offer important insights into underlying medical conditions.

Advanced Wound Care: Gain expertise in advanced wound care procedures to treat injuries, infections, and shell damage in Hawksbill turtles. These procedures include wound debridement, surgical repair, and topical therapies. See a specialist or veterinarian who specializes in reptiles for advice on advanced wound care techniques.

- Behavioral Monitoring and Enhancement:

Expert Hawksbill turtle caretakers can enhance their observing abilities and enrichment techniques to encourage organic behaviors and cognitive stimulation. Take into consideration these sophisticated methods:

Behavioral Observation: To learn more about the unique personalities, preferences, and communication signals of your Hawksbill turtles, set aside some time to observe their behavior, interactions, and social dynamics. Utilize this knowledge to customize habitat changes and enrichment activities to suit their individual requirements.

Environmental Enrichment: To keep your turtles cognitively and physically challenged, always be innovating and improving your environmental enrichment techniques. To maintain your turtles' interest and stimulation, provide new stimuli, alter the furniture in their environment, and try different feeding techniques.

Cognitive Enrichment: To test your Hawksbill turtles' cognitive capacities and encourage mental stimulation, use cognitive enrichment activities like puzzles, training exercises, and problem-solving exercises. Promote experimentation, curiosity, and adaptable behavior to improve their general wellbeing.

- Cooperation in Conservation and Research:

Expert Hawksbill turtle keepers can support scientific studies and conservation initiatives that safeguard and maintain wild populations. To aid in the conservation of Hawksbill turtles, think about taking part in cooperative research projects, citizen science campaigns, and conservation initiatives:

Research Collaboration: To provide important information, observations, and insights on the behavior, ecology, and captive care of Hawksbill turtles, collaborate with scientists, researchers, and conservation groups. Take part in field surveys, data collection projects, and research investigations to enhance scientific understanding and conservation initiatives.

Inform people on the value of protecting Hawksbill turtle populations and the challenges they face, including poaching, pollution, habitat loss, and climate change. In order to encourage conservation action and legislative change, raise awareness through public outreach, educational efforts, and advocacy campaigns.

Support habitat restoration efforts and activities that aim to restore important areas for Hawksbill turtles to nest, forage, and migrate. Take part in initiatives to preserve coral reefs, restore mangroves, and clean up beaches in order to enhance the resilience and quality of habitat for wild populations.

Expert Hawksbill turtle keepers may improve their husbandry methods, their turtles' wellbeing, and conservation efforts to preserve these amazing animals for future generations by putting these cutting-edge care suggestions and approaches into practice. It is important to always be learning new things, improving your abilities, and working together with other reptile enthusiasts to encourage appropriate handling and conservation of Hawksbill turtles, both in the wild and in captivity.

# Chapter 7

## Legal and Moral Issues with Hawksbill Turtle Residency

Responsible guardians of Hawksbill turtles must abide by significant ethical and legal requirements while keeping them as pets. Prioritizing the welfare of these endangered species and making sure that pertinent laws and ethical standards are followed are crucial. This includes knowing the rules governing turtle ownership and supporting conservation efforts. This thorough book will cover the legal requirements for owning Hawksbill turtles, the moral obligations of keepers, conservation initiatives, and strategies for encouraging the correct care of these amazing animals.

- The law governing the ownership of hawksbill turtles:

Understanding the laws pertaining to turtle ownership in your area is essential before obtaining a Hawksbill turtle as a pet. Country, state, and local laws and regulations pertaining to the ownership of turtles can include prohibitions on the importation, sale, and trade of endangered species. Think on the following legal facets:

Protection of Endangered Species: The International Union for Conservation of Nature (IUCN) has designated hawksbill turtles as critically endangered because of threats such as habitat loss, pollution, poaching, and other issues. To prevent the exploitation and trafficking of endangered species, such as Hawksbill turtles, numerous nations have passed laws and restrictions.

CITES Regulations: The Convention on International traffic in Endangered Species of Wild Fauna and Flora (CITES) forbids the international commercial traffic in specimens of the species, and hawksbill turtles are included on Appendix I of the list. The sale and import of

Hawksbill turtles and their goods across international borders is prohibited by CITES legislation.

National and Regional Laws: Find out what laws and rules apply to the ownership, possession, and trading of turtles in your nation or state on a national and regional level. Hawksbill turtle pet ownership may be completely prohibited in some areas, while others may call for permits, licenses, or registrations.

Wildlife Protection Organizations: Seek advice from organizations that safeguard wildlife, like the U.S. For information on legal requirements and compliance measures linked to ownership of Hawksbill turtles, contact the Department of Environment and Natural Resources (DENR) in the Philippines or the Fish and Wildlife Service (USFWS) in the United States.

- The Ethical Obligations of Keepers:

Keepers of Hawksbill turtles have an ethical obligation to protect the welfare and conservation of these endangered species in addition to adhering to the law. Among the ethical things turtle keepers should think about are:

Spread knowledge on the state of Hawksbill turtle conservation and the dangers that face natural populations, including habitat loss, pollution, climate change, and illicit commerce. Inform people of the value of defending threatened species and encouraging habitat conservation initiatives.

Responsible Ownership: Treat Hawksbill turtles with respect by giving them the proper attention, shelter, and enrichment to suit their needs on a behavioral, psychological, and physical level. Don't buy turtles from unethical or unsustainable sources, and don't contribute to the illegal trade in endangered species.

Rescue and Rehabilitation: Report sightings to wildlife officials and aid individuals who are injured or in distress in order to support rescue and rehabilitation efforts for sick, injured, or abandoned Hawksbill turtles. To help with the care and rehabilitation of rescued turtles, volunteer at animal rehabilitation facilities or with environmental organizations.

Action and Advocacy: At the local, state, and federal levels, push for more robust legal safeguards and conservation initiatives for Hawksbill turtles. Engage in lobbying efforts, sign petitions, and lend your support to legislative measures that would improve wildlife protection regulations and stop the illegal trade in endangered animals.

- Initiatives to Preserve Hawksbill Turtles:

In order to protect Hawksbill turtles and preserve their habitats for future generations, conservation activities are essential. Participate in conservation efforts and provide your support to groups whose mission it is to save Hawksbill turtles and their environments. Take a look at these conservation initiatives:

Protection of Nesting Beaches: To reduce disturbances and increase nesting success rates, support initiatives to safeguard and keep an eye on Hawksbill turtle nesting beaches, nesting locations, and hatchling emergence places. Participate as a volunteer in conservation initiatives that emphasize habitat restoration, predator control, and nest monitoring.

Marine Habitat Conservation: Promote the preservation and conservation of marine environments, such as seagrass beds, coral reefs, and coastal ecosystems, as these are vital to the survival and procreation of Hawksbill turtles. Encourage the creation of marine protected areas (MPAs) and other programs aimed at reducing risks from overfishing, pollution, and habitat loss.

Research and Monitoring: Participate in scientific studies focused on genetic diversity, reproductive biology, migration patterns, and populations of Hawksbill turtles. Take part in field surveys, citizen science initiatives, and data collection campaigns to provide important data for management and planning of conservation efforts.

Community Involvement: Involve regional communities, interested parties, and native populations in conservation initiatives and cooperative habitat management for Hawksbill turtles. To encourage sustainable practices and lessen human influences on turtle populations, cultivate partnerships with coastal communities, fishermen, and tourism operators.

- Encouraging Conscientious Stewardship:

Promoting appropriate stewardship among those who care for Hawksbill turtles is crucial to their welfare as well as to the protection of wild populations. Take into account the following strategies to encourage good stewardship:

Education & Outreach: Through educational programs, workshops, and outreach events, inform others about the proper maintenance and conscientious ownership of Hawksbill turtles. Disseminate information regarding food needs, habitat requirements, and ethical issues for turtle owners.

Adoption and Rehabilitation: If you see a reputable wildlife rehabilitation facility or conservation organization offering rescued or rehabilitated Hawksbill turtles for adoption, please consider doing so. Support centers offer sick, injured, orphaned turtles that cannot be returned to the wild a long-term care, treatment, and housing.

Sustainable Practices: To reduce environmental effects and advance conservation ideals, embrace eco-friendly substitutes and sustainable practices in daily life, habitat management, and turtle care. In order to preserve marine environments and species, cut back on plastic usage, recycle debris, and promote sustainable seafood options.

Cooperation and networking: Work together on conservation projects, share experiences, and exchange knowledge with other turtle caretakers, lovers, and conservationists. Become a member of social media groups, forums, and local or virtual turtle-keeping communities to meet like-minded people and support group conservation efforts.

Keepers of Hawksbill turtles can help ensure that these amazing animals are protected and preserved for future generations by incorporating ethical considerations, legal compliance, conservation initiatives, and responsible stewardship into turtle care methods. Never forget to put the needs of captive turtles first, raise public awareness of conservation issues, and push for more robust laws protecting Hawksbill turtles in the wild. When we work together, we can significantly improve the situation for the preservation of threatened species' natural areas.

# Chapter 8

## FAQs: Common Questions Addressed Concerning Hawksbill Turtles

People all across the world are fascinated by hawksbill turtles, which are amazing animals. As a result, many people have inquiries concerning their upkeep, conduct, state of conservation, and other topics. To help you learn more about these amazing reptiles and responsible care for them, we've included answers to some of the most commonly asked questions about Hawksbill turtles in this extensive guide.

A Hawksbill turtle: what is it?

The sea turtle species known as the Hawksbill (Eretmochelys imbricata) is distinguished by its remarkably exquisite shell, which has overlapping scutes that mimic a hawk's scales. Around the world, hawksbill turtles can be found in tropical and subtropical waters, mostly in coral reef environments. They get their name from their characteristic hooked beak, which they utilize to eat mostly sponges.

What is the Hawksbill turtle's state of conservation?

The International Union for Conservation of Nature (IUCN) has designated hawksbill turtles as critically endangered because of a number of concerns, such as habitat loss, pollution, climate change, illegal trafficking, and accidental catch in fishing gear. Since their population has drastically decreased over the past few decades, hawksbill turtles are now protected by international treaties as well as national regulations in many nations.

Can I have a pet Hawksbill turtle?

Because of their endangered status and legal restrictions, hawksbill turtles are prohibited from being kept as pets in the majority of countries. It can also be difficult to provide for the unique dietary and environmental needs of Hawksbill turtles when they are in captivity. Prioritizing the preservation of wild populations should be your top priority. Hawksbill turtles should not be kept as pets unless you are a professional wildlife rehabilitator or have the facilities and licenses necessary to properly care for them.

What nourishes the Hawksbill turtles?

Being mostly herbivorous, hawksbill turtles mostly eat the sponges that grow on coral reefs. They have been observed to eat a wide range of different invertebrates, algae, and plant materials, including seaweed, mollusks, sea anemones, and jellyfish. A range of leafy greens, veggies, fruits, and commercial turtle pellets designed for omnivorous turtles can be fed to Hawksbill turtles kept in captivity.

How large are mature Hawksbill turtles?

In comparison to other sea turtle species, hawksbill turtles are quite small; their average shell length is 2.5 to 3 feet (75 to 90 centimeters), and their usual weight is 100 to 150 pounds (45 to 68 kilograms). Individuals can differ in size, though, based on things like age, sex, and geography.

What is the lifespan of Hawksbill turtles?

Although the exact longevity of Hawksbill turtles in the wild is unknown, it is thought that they can live for several decades or longer. With the right maintenance and care, Hawksbill turtles can survive for up to 30 years in captivity.

Do aggressive Hawksbill turtles exist?

Although hawksbill turtles aren't usually thought of as hostile toward people, they can act defensively if they feel threatened or provoked. Hawksbill turtles should be handled carefully and with respect, just like any other natural species. Try to avoid causing them any stress or disruption by interacting with them as little as possible.

How can I support the preservation of Hawksbill turtles?

You can support the preservation of Hawksbill turtles and their habitats in a number of ways:

- Help conservation organizations: Contribute to respectable groups like the Marine Conservation Society, WWF, and Sea Turtle Conservancy that are devoted to preserving sea turtles and their ecosystems.

- Reduce plastic pollution in marine habitats, which can endanger Hawksbill turtles and other marine life. Cut down on your usage of single-use plastics and take part in beach clean-ups and coastal conservation initiatives.
- Adhere to responsible tourism guidelines: Pick environmentally conscious tour companies and travel locations that value sustainable tourism practices and contribute to conservation campaigns that save sea turtles and their nesting grounds.
- Report sightings and strandings: To support research and monitoring efforts, report Hawksbill turtle sightings, nesting activities, and strandings to local wildlife authorities or sea turtle conservation organizations.

How should I respond if I come across an injured or stranded Hawksbill turtle?

The following must be done right away in order to help a stranded or injured Hawksbill turtle and protect their wellbeing:

- Maintain a safe distance: Approach the turtle with caution so as not to stress or hurt it more. To lessen distractions, keep pets and spectators away from the turtle.
- Get in touch with the appropriate authorities: Request help from regional wildlife authorities, marine animal rescue groups, or sea turtle conservation organizations. Give details regarding the location, health, and presence of any obvious injuries on the turtle.
- If required, provide the turtle a makeshift home by putting it in a well-ventilated, shaded container and covering it with a damp towel or cloth to keep it from drying out while you wait for assistance to arrive.

- If you are not trained and permitted to handle or transport the turtle yourself, do not attempt to do so. To guarantee appropriate care and treatment, stranded or injured turtles should be treated by qualified wildlife rehabilitators or veterinarians.

After rehabilitation, can Hawksbill turtles be put back into the wild?

Hawksbill turtles may occasionally be allowed to return to the wild after being saved, treated, and found to be healthy by veterinarians or wildlife rehabilitators. Releasing decisions are contingent upon a number of factors, including as the health of the turtle, its capacity for wild survival, and the goals of local conservation efforts. To increase their chances of survival, rehabilitated turtles must go through extensive behavioral assessments, health examinations, and acclimation processes before being released.

Are items made from Hawksbill turtles subject to any regulations?

Yes, the international commerce of products made from Hawksbill turtles, such as their eggs, flesh, and shells, is subject to stringent laws. International conventions like the Convention on International Trading in Endangered Species of Wild Fauna and Flora (CITES), which forbids the commercial trading in specimens of the species, protect hawksbill turtles and their goods. Furthermore, a lot of nations have passed national legislation and rules to uphold these safeguards and stop the unlawful trade in endangered animals.

Is it possible for me to witness Hawksbill turtles on nesting beaches?

While visiting beaches where Hawksbill turtles nest can be an instructive and gratifying experience, it is imperative to do so in an ethical and responsible manner. In order to prevent disturbances to nesting turtles and their nests, access to many nesting beaches may be limited or regulated as they are situated within protected areas or marine reserves. Prior to visiting, learn about local laws, policies, and recommended procedures regarding turtle viewing. You should also follow any instructions issued by park officials or environmental groups.

By responding to these frequently asked questions, we hope to encourage a deeper comprehension, admiration, and preservation of these amazing animals. It's important to put the wellbeing of Hawksbill turtles first and support initiatives to save and maintain their habitats for coming generations, regardless of whether you come across them in the wild, see them at a conservation center, or think about keeping them in captivity.

# Chapter 9

## Final Thoughts: Savoring Your Hawksbill Turtle Friend

Well done for taking up the task of taking care of a Hawksbill turtle! It's crucial that you approach your job as a keeper with commitment, accountability, and a profound respect for these amazing creatures as you embark on this fulfilling journey. In this final segment, we'll highlight the most important lessons learned and offer suggestions for living a happy, meaningful life with your Hawksbill turtle companion while putting their protection and welfare first.

- Recognizing Your Keeper Role:

You are an essential part of a Hawksbill turtle's caretaker, meeting its physical, behavioral, and emotional needs. To make sure you can properly address the needs of Hawksbill turtles, it is imperative that you educate yourself about their natural history, habitat requirements, dietary preferences, and medical needs. Your turtle friend can thrive in a caring and stimulating environment that you establish by putting their welfare and well-being first.

- Establishing the Perfect Habitat

Creating a perfect home for your Hawksbill turtle is crucial to their general well-being and longevity. To replicate their native habitat, think about adding realistic features like driftwood, live plants, rocks, and substrate. Give them lots of room to swim, sunbathe, and explore. You should also offer opportunities for environmental enrichment to help them develop their physical and mental skills.

- Providing a Healthy Diet

Providing your Hawksbill turtle with a healthy, well-balanced diet is essential to preserving its vigor. To guarantee kids get all the vitamins, minerals, and other nutrients they need, offer a range of fresh foods, such as fruits, vegetables, leafy greens, and protein sources. To suit their unique dietary needs, add commercial turtle pellets and calcium supplements to their diet.

- Encouragement of Interaction and Enrichment:

Maintaining the mental stimulation and engagement of your Hawksbill turtle requires providing enrichment and interaction opportunities. To promote natural behaviors, exploration, and cerebral stimulation, provide an assortment of toys, puzzles, and environmental stimuli. To deepen your relationship and build mutual trust, spend quality time with your turtle through gentle touch, observation, and positive reinforcement.

- Setting Health and Well-Being First:

Sustaining the health and well-being of your Hawksbill turtle involves constant observation, prophylactic treatment, and swift action in the event of disease or damage. Plan routine vet visits, keep an eye on environmental factors, and watch your turtle's behavior and appetite for any indications of health problems. By practicing good husbandry and nutrition, you can proactively prevent common health issues including respiratory infections, shell rot, and metabolic bone disease.

- Promoting Conservation:

In order to secure the long-term survival and well-being of Hawksbill turtles, it is imperative that stewards of these turtles actively promote their conservation. Participate in research projects, donate to conservation organizations, and spread the word about the dangers posed to Hawksbill turtles and their habitats. Urge people to embrace environmentally friendly habits, cut down on plastic waste, and back initiatives to save endangered animals and marine environments.

- Savoring the Trip:

Above all, embrace the special relationship you have with your Hawksbill turtle buddy and enjoy the process of caring for them. Spend some time observing their habits, admiring their beauty, and marveling at how well they have adapted to living in the ocean. Enjoy every moment spent with them, whether you're exploring their environment, watching them soak up the sun, or just enjoying peaceful time together. Don't miss the chance to pick their brains about these amazing animals.

In summary, taking care of a Hawksbill turtle is both an honor and a duty that calls for devotion, empathy, and a shared commitment to environmental preservation. You can have a rewarding and fulfilling relationship with your Hawksbill turtle companion while supporting conservation efforts and helping to protect and preserve these amazing creatures for future generations by creating a nurturing environment, encouraging enrichment and interaction, placing a high priority on health and wellness, and speaking out in favor of conservation.